Project AIR FORCE

Supporting Expeditionary Aerospace Forces

NEW AGILE COMBAT SUPPORT POSTURES

Lionel Galway
Robert S. Tripp
Timothy L. Ramey
John G. Drew

Prepared for the
UNITED STATES AIR FORCE

RAND

The research reported here was sponsored by the United States Air Force under Contract F49642-96-C-0001. Further information may be obtained from the Strategic Planning Division, Directorate of Plans, Hq USAF.

Library of Congress Cataloging-in-Publication Data

Supporting expeditionary aerospace forces : new agile combat
support postures / Lionel Galway ... [et al.].
        p. cm.
      " MR-1075-AF."
    Includes bibliographical references.
    ISBN 0-8330-2801-4
      1. Space warfare. 2. United States. Air Force.—Commissariat.
I. Galway, Lionel A., 1950-

UG1530 .S87 2000
358' .8' 0973—dc21   99-089551

RAND is a nonprofit institution that helps improve policy and decisionmaking through research and analysis. RAND® is a registered trademark. RAND's publications do not necessarily reflect the opinions or policies of its research sponsors.

Published 2000 by RAND
1700 Main Street, P.O. Box 2138, Santa Monica, CA 90407-2138
1333 H St., N.W., Washington, D.C. 20005-4707
RAND URL: http://www.rand.org/
To order RAND documents or to obtain additional information,
contact Distribution Services: Telephone: (310) 451-7002;
Fax: (310) 451-6915; Internet: order@rand.org

This report addresses support of emerging Air Force employment strategies associated with Expeditionary Aerospace Forces (EAFs). EAF concepts turn on the premise that rapidly tailorable, quickly deployable, immediately employable, and highly effective air and space force packages can serve as a credible substitute for permanent forward presence. Success of the EAF will, to a great extent, depend on the effectiveness and efficiency of the Agile Combat Support (ACS) system in supporting expeditionary operations. This study is one of a series of RAND publications that address ACS issues in implementing the EAF. Others address planning, practices, policies, and technologies that can enhance the effectiveness of the EAF (see, e.g., Tripp et al., 1999).

Our initial work, documented here, shows that to implement the EAF concept the Air Force will need to develop a comprehensive system of forward support infrastructure. This work has been widely briefed within the Air Force at two meetings sponsored by the Air Force Deputy Chief of Staff for Installations and Logistics (AF/IL) and the Deputy Chief of Staff for Air and Space Operations (AF/XO), one meeting of the Logistics Board of Advisors, and several seminars held with functional experts on the Air Staff, as well as at Headquarters U.S. Air Forces Europe. Some of this material was presented by AF/IL at the CORONA meetings sponsored by the Air Force Chief of Staff. The research here is being extended and refined as part of the ongoing study of supporting expeditionary aerospace forces.

This research was conducted in the Force Modernization and Employment and Resource Management and Acquisition Programs of Project AIR FORCE as one element of a project entitled "Implementing an Effective Air Expeditionary Force." The project was sponsored jointly by AF/IL and AF/XO. This report should be of interest to logisticians and operators in the Air Force concerned with implementing the EAF concept.

Chief Master Sergeant John Drew is a staff member at the Air Force Logistics Management Agency.

PROJECT AIR FORCE

Project AIR FORCE, a division of RAND, is the Air Force Federally Funded Research and Development Center (FFRDC) for studies and analyses. It provides the Air Force with independent analyses of policy alternatives affecting the development, employment, combat readiness, and support of current and future aerospace forces. Research is performed in four programs: Aerospace Force Development; Manpower, Personnel, and Training; Resource Management; and Strategy and Doctrine.

CONTENTS

FIGURES

TABLES

INTRODUCTION AND MOTIVATION

With the end of the Cold War, the United States has entered an entirely new security environment in which the United States is the only global superpower in a world of many regional powers. The resulting demand for U.S. presence or intervention has required the U.S. Air Force to stage a large number of deployments, carried out by a substantially smaller force than existed in the 1980s. Whereas many of these operations required forward presence for patrol or combat (by rotating units from the continental U.S. [CONUS] for varying lengths of time), many others were and remain deterrent, in that forces are poised for a quick response but are not actively engaged (e.g., Phoenix Scorpion in Kuwait or Korea). The increased workload and operational turbulence have been blamed for a decrease in personnel retention and recent troubling decreases in overall readiness.

To meet the challenges of the new security environment with its uncertain demands and to address the problems of personnel turbulence, the Air Force has formulated the Expeditionary Aerospace Force (EAF) concept. The concept envisions highly capable and tailored force packages drawn from a set of Aerospace Expeditionary Forces (AEFs) that could be deployed from their home bases to provide air power on short notice anywhere around the world, which would allow greatly reducing (if not eliminating) deterrent deployments. The responsibility for responding would be rotated across AEFs, so that any one AEF would have a 90-day on-call period every 15 months. By using the AEFs to meet steady-state rotational deployments as well, the EAF concept should greatly decrease personnel turbulence and create more predictability in planning unit activities.

The shift to an Air Force structured for expeditionary operations (fast deployments to a breaking crisis, possibly to bases with minimal infrastructure in place) has presented the Air Force with a number of challenges in planning and support. In this report, we present analyses indicating that achieving the EAF goals will require strategic preparation of theater infrastructure: development of a global system of support locations (Forward Support Locations [FSLs] and in CONUS) that provide materiel, maintenance, and transportation to deployed units at Forward Operating Locations (FOLs). Determining how support activities are distributed among CONUS, FSLs, and FOLs is the essence of strategic support decisions. These locations must be connected by resupply (because FOLs and the deployed units will depend on CONUS and FSLs for some of their support) and will need to be coordinated by an enhanced logistics command and control (C2) system (because all support will not be with the unit).

GENERAL ANALYTIC FRAMEWORK

The primary challenge facing Air Force decisionmakers is uncertainty about almost every aspect of expeditionary operations: where will an expeditionary force be employed, when, and under what political conditions? Our approach to understanding such uncertainty is to build models of logistics support for various commodities, and run them under different scenarios to assess how logistics requirements change for various levels of conflict in different locations. Because our emphasis is on strategic decisions, the models need account only for major equipment and personnel.

Each of our models has three components. First is a mission analysis that specifies the critical mission parameters that determine the requirements for each support commodity. We then use employment-driven logistics process models, consisting of rules and algorithms, that compute the process timelines and requirements for material, equipment, and people to establish and operate the process. A third component evaluates the performance of alternative infrastructure options in providing these requirements (e.g., prepositioning all munitions at an FOL versus bringing in air-to-air missiles from CONUS).

We have developed models for munitions, fuel support, unit maintenance, vehicles, and shelters, and we have developed data for the models on the basis of extensive interviews at units that have deployed expeditionary forces to Southwest Asia (SWA). Further modeling work is planned to refine the current models and to develop methods of integrating the individual commodity models.

SUPPORT INFRASTRUCTURE COMPONENTS: FOLs AND FSLs

From this analysis, we have determined FOL characteristics and resource packages for a number of commodities to achieve different timelines for executing expeditionary operations. For analytic purposes, we have divided FOLs into three categories based on their infrastructure:

- A Category-3 FOL is the most austere FOL to which an expeditionary force would deploy; it meets only the minimum requirements for operation of a small fighter package (runway, water supply, fuel availability).

- A Category-2 FOL has the minimum requirements plus prepared space for fuel-storage facilities, a fuel-distribution system, general-purpose vehicles, and basic shelter.

- A Category-1 FOL has the attributes of a Category-2 FOL plus an aircraft arresting system, munitions buildup and storage sites already set up, and three days' worth of prepositioned munitions.

For each category of FOL, the resources that have not been prepositioned must be supplied during execution so that the supported force meets sortie-generation requirements. The options we consider for supplying these resources are supply from FSLs or from the CONUS.

EXPEDITIONARY DEPLOYMENT PERFORMANCE: PROTOTYPE ANALYSIS

How well can FOLs with varying amounts of prepositioned equipment support expeditionary operations in terms of timeline, airlift requirements, and cost? And what is the performance of the two options for supplying the materiel that is not prepositioned? The mission package we consider is a force that has been deployed in SWA: 12 F-15Cs, 12 F-16CJs, and 12 F-15Es accomplishing ground attack with 2000-lb precision munitions. For various categories of FOLs and additional resources from an FSL or from CONUS, we compute how long deployment takes, and a rough estimate of investment and recurring costs to support that option. Figure S.1 displays the timelines (to initial operational capability, or IOC) for various prepositioning options for each FOL category for this ground-attack operation. As expected, extensive prepositioning decreases the time required to deploy an expeditionary force, but it imposes a cost penalty for having large stocks of equipment in many forward locations.

Note that these timelines are functions of current support processes and practices. Modifications to these processes will reduce the timelines, and our analysis framework allows us to assess the potential payoff along several metrics. For example, lightweight munitions could reduce the deployment footprint and make airlift of ground-attack munitions feasible.

After examining today's force structure and its support processes, our analysis leads to a number of insights about FOLs and their support:

- To get close to the 48-hour deadline from execution order to dropping the first bombs on target, fighter expeditionary forces must deploy to Category-1 FOLs. Further, given that a flight halfway around the world takes approximately 20 hours, pushing the timeline below 48 hours will require having people permanently deployed, or materiel at an advanced state of preparation at the FOL, or both.

- Equipping many FOLs from scratch would be expensive. Although much of the cost for some materiel might well be sunk, maintenance and storage costs will still have to be paid.

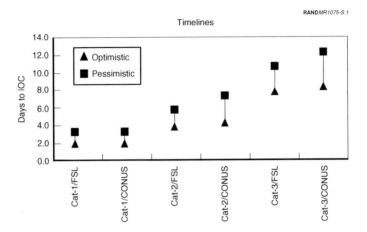

Figure S.1—Timelines for Infrastructure Options

- FSLs provide a compromise in cost between prepositioning at FOLs and deploying everything from CONUS. In terms of timeline, the FSL option offers only a slight advantage over the CONUS option (primarily because ramp space is an airlift limiting factor). However, airlift from an FSL does eliminate having to deploy a tanker air bridge from CONUS for support provided by strategic lift, leaving these resources available for other uses.

- Category-2 FOLs represent another compromise in cost and timeline. Deploying to a Category-2 FOL takes approximately three additional days to unload the airlift and two to three days to set up munitions and petroleum/oil/lubricants (POL) storage. There is also a risk that arrangements for rental vehicles, shelter, medical facilities, and the like that are anticipated to be available (perhaps commercially) would require additional time to finalize.

- Category-3 FOLs are not useful as FOLs for very quick crisis response, given the time required to unload airlift and set up support processes. However, this is a function of the current processes, and the timeline estimated here is for a stressing scenario. A less-stressing scenario or a humanitarian operation requiring simple support might well be feasible from a Category-3 FOL.

ANALYZING OPTIONS FOR EXPEDITIONARY ACS

As originally envisioned, the EAF would consist of AEFs that would provide air and space forces that could be tailored to emerging crises, deployed rapidly to the required location, and be prepared to execute operations immediately (within 48 hours of departure from home). The prototype analysis indicates that this is not possible: *With today's support processes, policies, and technologies, deploying even a modest fighter-based combat force to a bare base will require several days of development before the FOL can sustain a high flying tempo.*

This finding does not mean that achieving the 48-hour operational goal is impossible. The goal can be met by developing a strategic theater infrastructure with the judicious prepositioning of equipment, materiel, and facilities, but it requires hard thinking about threats and the level of U.S. interests involved to ensure that such investment is worth the cost. Other options include accepting a longer timeline, using bomber operations as a first response, and changing the current processes to add improved technology or new policies.

We assert that the long-term support issues raised in this report about FOLs, FSLs, and their locations and equipage require analyses carried out with a *strategic* perspective, one that views the entire support structure, both inside and outside CONUS, as a *system* of global support.

ACKNOWLEDGMENTS

Numerous persons inside and outside the Air Force have provided valuable assistance and support to our work. We thank General John Jumper for initiating this study when he was Deputy Chief of Staff, Operations (AF/XO), and, for his sponsorship and support, Lieutenant General William Hallin when he was Deputy Chief of Staff, Installations and Logistics (AF/IL). General Patrick Gamble, PACAF/CC, has been a staunch proponent of this effort, as well as Lieutenant Generals Esmond, AF/XO, and Handy, AF/IL. Dr. Robert Wolff, when he was AF/ILX, provided valuable guidance and assistance throughout the course of this research. Ms. Sue O'Neal, AF/ILX, has continued to support and sponsor this research.

At the Air Staff, we thank Major General Zettler and Mr. Dunn, AF/ILM; Brigadier General Stuart and Brigadier General (sel) Totsch, AF/ILS; Brigadier General Petersen, AF/ILT; and their staff for their support and critique of this work. We have also enjoyed support for our research from the Air Force's Major Commands that are responsible for implementing the EAF. Major General Dennis Haines, ACC/XR, and Brigadier General Terry Gabreski, USAFE/LG, and their staffs provided access to personnel and data at their operating locations.

Colonel Michael Weitzel, CENTAF/LG, and his staff, particularly Major Philip Moore (ret.), CENTAF/LGXX, helped in arranging visits to Southwest Asia, answering numerous questions and data calls on logistics operations and providing keen insights into the area of responsibility.

Several AEF deploying units provided complete access to deployment and employment processes and data. These included the 4th Fight Wing (FW) at Seymour-Johnson Air Force Base, the 20th FW at Shaw Air Force Base, the 169th FW at McEntire Field, and the 366th FW at Mountain Home Air Force Base. At each unit, the logistics staff provided us with the extensive interviews and data that made our modeling efforts possible.

Our research has been a team effort with the AF Logistics Management Agency (AFLMA); the support of the AFLMA has been critical to the conduct of this research. We wish especially to thank Colonel Richard Bereit, AFLMA/CC. Randy King of the Logistics Management Institute (LMI) has contributed to our analysis of spare parts. Colonel Donald Okrup and the personnel of the AEF Battle Lab have been helpful in collecting data and validating our models.

The authors wish to thank RAND colleagues Mort Berman and Marc Robbins for their thoughtful and constructive reviews, which contributed greatly to the clarity of the report.

Finally, we thank our action officers, Lieutenant Colonel Tony Dronkers, AF/ILXX, and Major Ernie Eannarino, AF/XOCD, for their encouragement and support.

ACC	Air Combat Command
ACS	Agile Combat Support
AEF	Aerospace Expeditionary Force
AEW	Air Expeditionary Wing
AFB	Air Force Base
AF/IL	Air Force Deputy Chief of Staff, Installations and Logistics
AFLMA	Air Force Logistics Management Agency
AF/XO	Air Force Deputy Chief of Staff, Air and Space Operations
AGE	Aerospace Ground Equipment
AOR	Area of Responsibility
ASETF	Aerospace Expeditionary Task Force
BOT	Bombs on Target
C2	Command and Control
CENTAF	Central Air Force
CINC	Commander in Chief
COB	Collocated Operating Base
CONUS	Continental United States
CSAF	Chief of Staff of the Air Force
DBOF	Defense Business Operating Fund
DPG	Defense Planning Guidance
EAF	Expeditionary Aerospace Force
ECS	Expeditionary Combat Support
FOL	Forward Operating Location
FSL	Forward Support Location
IOC	Initial Operating Capability

LANTIRN	Low-Altitude Navigation and Targeting Infrared for Night
LIN	Liquid Nitrogen
LOX	Liquid Oxygen
MOB	Main Operating Base
MOG	Maximum on ground
MDS	Mission Design Series
MTW	Major Theater War
PAA	Primary Aircraft Authorized
PGM	Precision-Guided Munition
POL	Petroleum/Oil/Lubricants
RSP	Resource Spares Package
SBS	Small Bomb System
SEAD	Suppression of Enemy Air Defenses
SWA	Southwest Asia
TALCE	Tanker Airlift Control Element
TCTO	Time Compliance Technical Order
TPFDD	Time-Phased Force Deployment Database
TWCF	Transportation Working Capital Fund
USAFE	U.S. Air Force Europe
WRM	War Reserve Materiel

Chapter One

THE NEW SECURITY ENVIRONMENT AND THE USAF

When the Cold War ended in 1989, symbolized by the destruction of the Berlin Wall, the United States entered a markedly changed security environment. In the course of a few years, that environment had changed from a bipolar world in which two superpowers confronted each other around the globe to a world in which the United States is the only global superpower among many regional powers and many regional conflicts. The resulting demand for U.S. presence or intervention has required deployments ranging in size and purpose from Operation Desert Storm and Operation Allied Force, through Northern and Southern Watch and Uphold Democracy,[1] to humanitarian relief and noncombatant evacuation operations.

The Air Force has played a large role in these operations, and the pace of its activity has not abated: Figure 1.1 illustrates the range of deployments the Air Force has faced in the 1990s (before Operation Allied Force in Kosovo). Not only are the operations far-flung, but many were initiated with short lead times in response to potential crises. Many of these operations involved patrol or combat (e.g., Desert Storm, Allied Force, Northern and Southern Watch), whereas many others were and remain deterrent in nature—U.S. forces are

---

[1]Operation Desert Storm was the U.S.-led war to evict Iraq from Kuwait. Operations Northern and Southern Watch are allied operations to prevent Iraq from making military flights in U.N.-designated no-fly zones. Operation Uphold Democracy reinstalled Haitian president Aristide by removing a military junta.

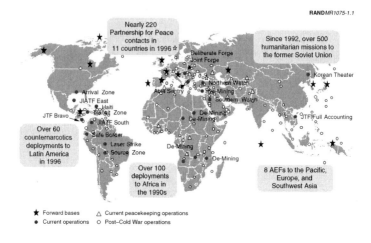

RAND*MR*1075-1.1

Figure 1.1—Recent USAF Deployments

stationed in critical regions to ensure a quick response if conflict occurs but are not actively engaged in combat or patrol missions. Note that even in Southwest Asia (SWA) there have been deployments such as Phoenix Scorpion whose primary purpose was to increase force levels in response to sudden Iraqi troop movements.

The number, frequency, and uncertainty of these deployments have created a number of problems for the Air Force. The deployments in the latter half of the 1990s are being carried out by a substantially smaller force than existed in the 1980s or even during Desert Storm. This has resulted in personnel turbulence, as specialists in critical fields are frequently sent on lengthy deployments, and increased workload, both for the deployed personnel and for the people left behind who must cover the home-base workload of those deployed. This turbulence has been blamed for a decrease in retention,[2] which,

---

[2]See, for example, Paul Richter (November 22, 1998). However, other research has shown that some deployment may improve retention. See Hosek and Totten (1998).

coupled with continually declining defense expenditures, has been linked by some to recent troubling decreases in overall readiness.[3]

THE EXPEDITIONARY AEROSPACE FORCE

To meet the challenges of the new security environment, the Air Force has formulated the Expeditionary Aerospace Force (EAF) concept.[4] Under this concept, the response to a fast-breaking crisis is to deploy a tailored air power force rapidly to the crisis area from bases primarily in the continental U.S. (CONUS), in contrast to the previous posture where forces were deployed overseas in areas of concern for lengthy periods as deterrents or in anticipation of crisis situations. The ability to respond rapidly to a crisis[5] from CONUS would greatly reduce the need to have forces deployed in critical areas purely for deterrence; such forces should also be more flexible in that they could deploy to any crisis area without first withdrawing from a current deployment, an action that often has political implications and restrictions.[6]

The EAF concept addresses many of the problems currently experienced. First, keeping most units in CONUS except when deployed

---

[3]Again, this has been the subject of many news stories. See Paul Richter (November 17, 1998) or Matthew Williams (September 28, 1998).

[4]See Davis (1998) for a description of the origin and early development of the EAF idea. Davis notes that the USAF has had "expeditionary" units before, notably the 19th Air Force, which was a headquarters unit designed to move quickly into theaters of operation and start employing arriving air units. One-third of the members of the 19th were jump-qualified.

[5]Rapidly deployable air power could be critical even in larger conflicts such as Major Theater Wars (MTWs): as a result of the Quadrennial Defense Review (QDR), analysts at RAND and the Air Staff pointed out that the effect of air power was being systematically undervalued in some combat assessments of potential MTW force mixes. The rapid deployment of air power, they asserted, could be used to blunt major armored attacks (such as occurred during the seizure of Kuwait by Iraq in 1990) without the necessity of engaging the enemy with major formations of friendly ground forces (Ochmanek, Harshberger, and Thaler, unpublished RAND research). However, to be effective halt phase (or effect-based) operations would require very rapid deployment and immediate employment, as well as a demanding operations tempo for the first few days of the battle.

[6]In one instance, Saudi Arabia temporarily blocked U.S. attempts to move U.S. aircraft within the country to other SWA locations in preparation for operations against Iraq. See Washington Post (February 3, 1998).

for a crisis situation should greatly decrease the extended deployments for deterrent purposes that are partially blamed for the retention problem. Second, rotating deployment responsibilities among units (so that each unit is on-call for a specified period of time) would create more predictability in planning unit activities such as training and periodic maintenance. The reduction in uncertainty about sudden deployments would also increase the quality of life for personnel. Finally, using some of the on-call units to staff the rotation requirements would even out these burdens.

These considerations led the Chief of Staff of the Air Force (CSAF) to hold a press conference in August 1998 at which he announced that the Air Force was adopting the EAF concept as its basis for responding to small-scale contingencies and staffing rotations and described the framework for moving the force to an expeditionary posture.[7] Emphasizing that there were many more details to be decided, he outlined the division of the active forces into 10 Air (later "Aerospace") Expeditionary Forces (AEFs),[8] each a mixture of fighters, bombers, and tankers. This organization aligns forces from the current predominantly single-MDS (mission/design/series) wing structure with the 10 AEFs; it is to be operational by January 2000. Two of the AEFs will be on call for a 90-day period, when elements from those AEFs could be deployed for any crisis needing air power. The on-call period will be followed by a 12-month period during

---

[7] Press conference August 4, 1998, at the Pentagon, held by Acting Secretary of the Air Force F. Whitten Peters and CSAF Gen Ryan. This is the most comprehensive of several talks on the subject by Gen Ryan. See Ryan (1998).

[8] There is a terminology problem in discussing the structure and employment of expeditionary forces: although the basic concept has remained the same, the name has gone through several iterations. The original expeditionary force package, tailored to SWA, was a 30- or 36-ship fighter package, which was termed an Air Expeditionary Force (AEF). The concept was broadened to include other types of missions, including humanitarian and space support (hence the replacement of "Air" by "Aerospace"). Finally, as it became clear that this would be a significant shift in Air Force culture, the new organizational concept was named the EAF, which has been implemented by dividing the Air Force into AEFs as noted in the text. The generic term for the force package actually employed is Aerospace Expeditionary Task Force (ASETF), which we will use here, although AES (for squadrons), AEW (for wings), and AEG (for "groups") have been used. We will use EAF and AEF as defined in the CSAF briefing. See *Air Force Glossary*, AFDD 1-2, July 1999.

which the unit will carry out normal training activities.[9]   The deployed forces will be tailored in size and/or capability to match the requirements of the situation. The details of the implementation are expected to evolve.

As the EAF concept has been developed, it has become clear that the move to an Expeditionary Aerospace Force will require extensive re-thinking and reengineering of most of the Air Force; some have described it as a "cultural shift." Whereas the previous focus of the United States was on the European theater against the Warsaw Pact, and, to a lesser extent, on war on the Korean peninsula, now there is uncertainty in location, uncertainty about the enemy, and uncertainty about intensity, duration, and forces. The operational challenges to such a mode of fighting are many: obtaining intelligence information, formulating target lists, and devising operational plans all while deploying in preparation for going into action on arrival. In addition, the USAF had largely planned on going to war by deploying to bases with a large U.S. presence in place. The assumption was that only aircraft and personnel would be deployed, and those units would fall in on well-equipped bases. As a result of this focus on fighting from established bases, existing support equipment is heavy and not easily transportable, so that deploying all the support for almost any sized ASETF to an overseas location would be expensive in both time and airlift. In contrast, expeditionary deployments might be made to areas with little or no U.S. presence (such as was the situation at the beginning of Desert Shield). Indeed, the initial statements of the EAF concept talked about operations and supporting a force from austere bases with little if any infrastructure other than a usable runway.

Adapting Agile Combat Support (ACS) to be Expeditionary Combat Support (ECS) is therefore one of the greatest challenges posed by expeditionary operations. In an expeditionary world, agility requires that support processes be capable of supporting rapidly deployed forces, either by deploying rapidly themselves or by connecting support processes in permanent locations to the deployed forces.

---

[9]In the case of an MTW, on-call forces might well be used for a first, fast response, but forces in other parts of the cycle will be called as needed to carry out their various MTW missions.

This reshaping of support in implementing the EAF has been the focus of our research.

EXPEDITIONARY COMBAT SUPPORT

The Air Force faces a significant challenge in adapting support to a world of short-notice deployments, expeditionary operations, and fast-breaking theater conflicts. Much of the effort so far has focused on the logistics aspects of execution, i.e., how to compress the time required to deploy a unit's logistics support, given current processes and equipment. The Air Force has made progress in that area, as can be seen in Figure 1.2.

Figure 1.2 illustrates the current Air Combat Command (ACC) standard for deployment: 72 hours of strategic warning, followed by 24 hours to start the deployment, followed by another 18–24 hours to arrive in the theater, prepare the aircraft for combat, and begin to launch strikes. The ovals list the execution tasks to be accomplished

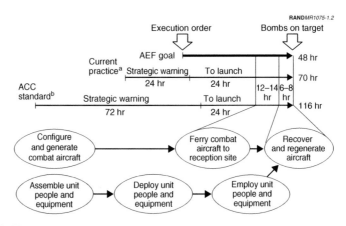

a AEF 4 experience.
b For 24 PAA units (AFI 90–201, ACC SUP 1, 1 Jan 96).

Figure 1.2—Deployment Timelines

when strategic warning is given. "AEF 4" (the 4th Fighter Wing's short-notice deployment to Qatar)[10] made substantial improve-ment on that timeline, but the goal for the EAF is more stringent still.

Rather than addressing execution, our research concentrates on the *strategic* decisions that will design the logistics infrastructure to support rapid deployments. Figure 1.3 illustrates the relationship of strategic decisions to the execution decisions listed in Figure 1.2.

The ovals indicate areas outside the deployment/redeployment execution that need to be addressed. Many of these areas are topics of ongoing RAND research: logistics command and control (C2), preparation of deploying units, policies for preparing airlift and

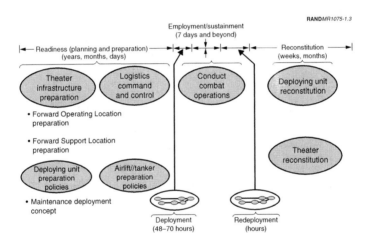

Figure 1.3—Strategic ACS Decisions

---

[10]Since the advent of the EAF concept, several of the deployments to SWA for rotational duties have been termed "expeditionary" and were numbered in sequence. In AEF 4 the actual start order for the deployment was given on short notice, although the wing knew that it was scheduled to deploy and had undergone preparation. The current usage of AEF is more general.

tankers, and preparation of theater infrastructure. In addition, re-constitution requirements need to be considered in each of the areas. We concentrate on theater infrastructure, but first briefly describe the other strategic decision areas.

Logistics C2 is the planning and coordination of ASETF logistics support, including selecting units and the deployment bases, determining the capabilities of the base, and coordinating the deployment. The urgency of the deployments requires that expeditionary logistics C2 be much faster and much more closely linked to operations than is the current practice.

Deploying-unit preparation is the coordination of unit preparation activities so that an ASETF can deploy quickly and operate with minimal overhead for an initial time if it is called during its on-call period. In aircraft maintenance, for example, preparation for an ASETF may include finishing required periodic inspections and modifications (Time Compliance Technical Orders, or TCTOs), and changeout of engines so that little scheduled maintenance will be needed at the base. Personnel preparation would include field training for austere billeting, small-arms training, and medical preparations. Airlift/tanker preparations might include streamlining the process of developing the deployment schedule (Time-Phased Force Deployment Database, or TPFDD), investing in tanker bases, prepositioning tanker and airlift assets and crews on strategic warning, and collocating airlift assets with first-deploying units such as security police for force protection.

THEATER INFRASTRUCTURE PREPARATION

In the original EAF concept, ASETFs would be able to deploy to any airfield around the world that had a runway capable of landing the operational and airlift aircraft. Some such airfields are fully equipped military bases; others could be "bare bases," with a minimum of infrastructure. To be flexible and truly expeditionary, an ASETF's reliance on prepositioned equipment and materiel was to be minimized if not eliminated.

However, analysis by RAND (see Chapter Four) and others[11] shows that prepositioned equipment and supplies cannot be eliminated; current logistics processes and equipment are simply too complex and heavy to deploy rapidly. New technologies and policies can improve this situation in the mid to long term, but implementing the EAF over the next few years will require some judicious prepositioning. Providing Agile Combat Support for the EAF today requires an expeditionary basing structure for support.

To analyze the required basing structure, we first developed the analytical framework described in Chapter Two. This framework uses a series of models that evaluate the major logistics processes in terms of their airlift requirements, time to move and set up, and cost. We address five major categories of resources: munitions, POL support (part of base support), unit maintenance equipment (the bulk of unit support equipment), vehicles, and shelter. These five commodities make up the majority of support materiel for an air operation. Figure 1.4 shows what these proportions were for the 4th Fighter Wing's deployment to Qatar; other deployments had similar patterns.

This recent deployment did not take place on very short notice nor was there substantial reengineering to tailor support processes and equipment. Our models represent the individual processes in enough detail so they can be used to evaluate such process modifications.

Chapter Three shows that to meet the demanding timelines and operating tempos (optempos) some thought must go in to building up infrastructure in all theaters where ASETFs could be deployed. Even though access to much or all of the infrastructure would be under the control of a foreign government, some risks must be taken to meet the deployment and employment goals.

Decisions about what and where to preposition are the basis of infrastructure preparation. Tradeoffs among a number of competing objectives must be analyzed: timeline, cost, footprint, risk, and flexibility. Prepositioning everything at the base from which operations

_____

[11]"Bare Base Analysis," briefing presented by AF/ILXX to Lt Gen Handy, AF/IL, on May 10, 1998. The briefing summarizes work done by ILXX and Synergy, Inc.

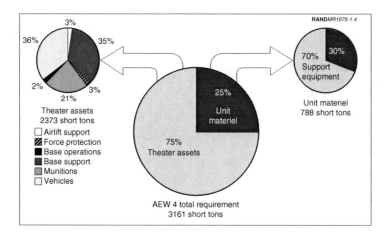

Figure 1.4—Breakdown of Support for AEF 4

will be flown reduces to a minimum the timeline and required airlift, but it also reduces flexibility, adds political and military risk, and incurs a substantial cost if several such bases are to be prepared. Bringing support from CONUS or a theater support location increases flexibility and reduces risk, but takes longer and requires more airlift.

Before describing in the next three chapters a methodology that allows us to assess how different configurations perform and analyze the tradeoffs in designing a forward infrastructure that meets Air Force needs, we briefly define the four basic components of forward infrastructure. First is the Forward Operating Location (FOL) from which the aircraft fly their missions. Many types of airfields are suitable for air operations; the different types of FOLs will be described more fully in Chapter Three. Each category of FOL requires different amounts of equipment to be brought in to prepare the FOL for operations and therefore has a different timeline and transportation requirement. One key decision about theater infrastructure is how many FOLs of each type the Air Force needs in a critical area and whether the United States will commit to equipping them with prepositioned equipment if needed to make deploying an ASETF easier.

The second component is the Forward Support Location (FSL), a support facility outside of CONUS but not (necessarily) in a crisis area. FSLs can be depots for U.S. war reserve materiel (WRM) storage, for repair of selected avionics or engines, a transportation hub, or a combination thereof. An FSL could be manned permanently by U.S. military or host-nation nationals, or simply be a warehouse operation until activated. The exact capability of an FSL will be determined by the forces it will potentially support and by the risks and costs of positioning specific capabilities at its location.

The third and fourth components are assured transportation/resupply and logistics C2. If ASETFs must deploy with minimum support and depend on resupply from either CONUS or a set of FSLs, they will need to have an assured resupply link whose responsiveness aligns with the support that is available at the FOL. If they deploy to a truly bare base, resupply will be required to keep vital supplies such as munitions flowing. Without the needed assured resupply, ASETFs will not be able to carry out their missions.[12] The strategic infrastructure envisioned here will require a more sophisticated support C2 structure to coordinate support activities across FOLs and FSLs connected by a rapid transportation system. These last two components are the subject of current RAND research.

The theater infrastructure is a combination of FOLs, FSLs, and assured resupply. Our contribution here is to provide tools and a pro-

---

[12]Assured transportation or resupply has been a contentious issue between deploying forces and theater Commanders-in-Chief (CINCs). If ASETFs increase their deployment speed by relying on support functions at FSLs or in CONUS, they will require frequent resupply to remain effective. As we will show below, prepositioning complete sets of supplies at numerous FOLs is probably not economically feasible and may be politically risky. However, when a crisis requires the rapid deployment of ground troops, the airlift requirement of the latter far exceeds the amount required even by large air units. Further, by doctrine the CINC has complete control of transportation into the theater, and partially deployed fighter wings did have resupply cut off for a time in Desert Shield. Although it would be prudent to have redundant transportation links from FSLs to FOLs, theater transportation infrastructure may limit their availability. Ultimately, the EAF concept requires a thorough understanding of what the transportation needs are for ASETFs and what would be the impact of interrupting that resupply flow.

totype of the analysis and planning that we believe the Air Force must do to prepare to deploy quickly to crisis spots around the world. The results are not definitive but should provide a starting point for evaluating alternative forward infrastructures for any theater and a wide range of ASETF deployments.

Chapter Two

GENERAL ANALYTIC FRAMEWORK

To reiterate: The goal of our analysis is to assist in making strategic decisions about forward logistics infrastructure—what and how much should be prepositioned at an FOL, what capabilities should be established at FSLs, and what should be deployed from CONUS. The primary challenge facing Air Force decisionmakers is uncertainty about almost every aspect of expeditionary operations: Where will an ASETF be employed, when, and under what political conditions. Our approach to understanding such uncertainty is to build models of logistics support for different commodities, and to use them to assess how logistics requirements change under different situations. Because our emphasis is on strategic decisions, the models need not be extremely fine-grained: as long as they account for major equipment and personnel, they will be adequate to assess strategies of FOL/FSL/CONUS tradeoffs. However, they must be detailed enough to help evaluate the effects of process improvements resulting from new policies or technologies, and to indicate which resources are constraints if the logistics requirements exceed the capabilities of logistics processes.

DETAILED DESCRIPTION

Our modeling approach is illustrated in Figure 2.1. The left panel represents the mission analysis. We extract the mission parameters that determine the requirements for each commodity. For munitions, for example, these parameters are the aircraft types, their missions (dictating the munitions they carry and expenditure rates), and their flying schedule (closely spaced launches require more loaders

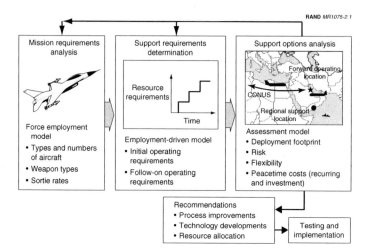

Figure 2.1—Analytic Framework

and possibly more facilities and personnel for buildup). In contrast, a model to determine shelter or services requirements would primarily be driven by the number of personnel deployed.

The middle panel represents the employment-driven process model. Here, the mission parameters are used to determine the support resources needed for each commodity. The munitions model derives the numbers of munitions expended daily from the flying schedule and expenditure rates, and then uses the daily expenditures to compute the equipment and personnel needed for munitions storage, buildup, and loading. For some commodities, this requirement can be time-phased in terms of either optempo or initial versus full operating capability of the support process. Most of the models were built as simple spreadsheets because the spreadsheet paradigm allows us to lay out the structure of the model and its data clearly and openly to users.

The right-hand panel assesses options for providing the resources computed in the second phase. For infrastructure decisions, the options are combinations of FOLs (of varying categories), FSLs, and support direct from CONUS, and the option analysis computes the time before a given process can support operations, the necessary transportation resources, and the costs for investment and operation. In the munitions model, for example, one option is to preposition three days' worth of munitions at each FOL and provide the rest from FSLs in the theater. Given a specific scenario, the model can compute the cost of each of these stockpiles as an investment cost for setting up this configuration.

Note in Figure 2.1 the top arrows feeding back to the mission analysis model both from the process model and from the assessment model. Some operational plans may be infeasible in terms of the deployed support process or in terms of possible forward infrastructure options, and a decision will need to be made whether to modify the operational requirements or the support options. The results of the model analyses will be recommendations for forward infrastructure, as well as changes in policies and technologies, all of which will be candidates for testing by the Air Force.

MODELS AND CURRENT STATUS

The results in Chapter Four derive from the models currently in use. Three of the models have been developed and are in the process of being extended and documented. The others are under development but have substantial representations of their commodities. However, there are a few other commodities (such as engines and their maintenance) that must still be addressed, which is why we regard our results as prototypical.

The munitions model is the most advanced of the models to date. Its mission inputs are aircraft type, mission, and loading schedule. Using aircraft loadouts, expenditure factors, and munitions characteristics, the model computes personnel and equipment requirements for the entire process from receipt and storage through loading of munitions on the aircraft. The model also computes the total amount of munitions used, munitions cost, and the airlift required to move different portions (e.g., bombs, missiles, and support equipment). Infrastructure options include prepositioning at FOLs (initial stockpiles versus full stockpiles) and resupply from FSLs or CONUS.

The POL support model is similar to the munitions model in its inputs and outputs, although here the modeling is largely "basic physics," depending on aircraft and storage capacities and flow rates.

The Air Force Logistics Management Agency (AFLMA) has developed a series of spreadsheet models that compute maintenance personnel and unit equipment for different MDS, based on in-depth interviews with maintainers at active units. An F-15E model has been largely completed, with models for the F-15C, F-16CJ, KC-135, B-1B, and B-52 in various stages of completion.

Initial work has been started on spreadsheet models for vehicles and shelter. The former consists of a list of vehicles needed for a fighter operation derived from focus groups convened at ACC and the Air Staff. Currently, we have firm numbers only for a seven-day ASETF package of fighters, although we also have some information on the vehicles required for bomber operations. The shelter model is largely a listing of the contents of Harvest Falcon (the current Air Force equipment set that provides shelter, water, power, kitchen, and work enclosures for operations at a bare base), although that data have also been used to sketch out a version of a proposed Harvest Phoenix (very austere billeting) equipment set.

Additional work supporting deployment packages and forward infrastructure option assessment has been done on resource spares package (RSP) kits and on repair of F-15 avionics, electronic warfare (EW) pods, and LANTIRN (Low-Altitude Navigation and Targeting Infrared for Night) pods. There are still other functions affecting deployment that we have not studied and are not included in this initial analysis. For example, force protection is a function of the security environment of the FOL and setup times could vary widely for different types of threats.

DATA SOURCES

In building these models, our primary source of information was interviews in the field with units that had deployed with AEFs to SWA, supplemented by interviews at the Air Staff and CENTAF (Central Command Air Force, the air component responsible for air operations in SWA) and by statements of policy in Air Force publications. We also interviewed functional experts in other units and such organizations as the Air Force Combat Ammunition Center (AFCOMAC). The goal was to estimate the minimum required to staff and support each process. We were most interested in using our models as a general framework for analysis, one in which rules and data could be examined, rigorously questioned, and modified according to experience, testing, and changing circumstances. The data are therefore realistic, coming from actual units that have made this type of deployment several times. In some cases where data were difficult to obtain, we made educated guesses and carefully documented the rationale so that users could judge the adequacy of the inputs.

INTEGRATING MODELS

The models described above focus on single commodities, although they do cut across organizational lines where necessary (e.g., the munitions support model covers both munitions buildup and aircraft loading processes). Eventually all of the resources must be brought together in a deployment. To integrate the process models, we are developing an optimization model to bring together the processes and help determine how to provide each segment of support. That work is not described here. Instead, we use a simpler integration (described in Chapter Four) of the support processes to help assess forward infrastructure options.

Chapter Three

SUPPORT INFRASTRUCTURE COMPONENTS:
FOLs AND FSLs

We next describe two major components of the support infrastructure—FOLs and FSLs. As we shall see in the next chapter, the performance of these components in supporting expeditionary operations depends on the amount of materiel prepositioned at FOLs, as well as the options, such as a forward support location (FSL) near the theater or the CONUS, for supplying the remainder that is not prepositioned. We do not explicitly discuss the resupply and logistics C2 systems here, but it will be evident from the discussion that the FOL/FSL system depends heavily on both.

CATEGORIES OF FOL

To fight Cold War conflicts in Europe or on the Korean peninsula, the Air Force had planned to deploy massive amounts of air power to fixed FOLs, either to main operating bases (MOBs) already operated by the United States or to collocated operating bases (COBs) where a friendly air force was operating. In either case, the bulk of the deployment was aircraft and people; fuel, munitions, and most of the necessary support equipment and supplies were to be in place or moved to the FOL from storage locations.

With the new expeditionary focus, the Air Force must now consider deploying air power assets to a number of locations, each with differing characteristics. Clearly, the amount of equipment that is prepositioned can affect the timeline of deployment, the cost of setting up and operating a system of FOLs in peacetime, and the flexibility and risks of depending on the use of those FOLs in times of crisis. For analysis purposes, we define three categories of FOL. [1]

Category 3

A Category-3 FOL is the most austere FOL to which an ASETF would deploy. It has the minimum requirements for operation of a small fighter package—a minimum runway length of 8000 feet, a Load Classification Number (LCN) of 72 to handle C-141s,[2] a working Maximum on Ground (MOG) of two (narrow-body airlifters), sufficient water to provide 8 gal/day for 1100 people,[3] fuel supply of 158,000 gal/day, availability of LOX/LIN (liquid oxygen and liquid nitrogen), plus planned siting for munitions facilities. (This level of detailed knowledge implies that the field has been recently surveyed and that preplanning has indicated where facilities can be laid out.) For a Category-3 FOL, the Air Force must provide much of the airfield infrastructure, shelter for personnel, all munitions, fuel storage and distribution, and all vehicles.

Category 2

A Category-2 FOL is less austere than a Category-3 FOL in that it has prepared space for fuel-storage facilities (500,000 gal), a fuel-distribution system in place (e.g., refueling trucks), general-purpose vehicles for rental as well as fire and medical vehicles, medical facilities that the United States can use, and sufficient shelter for personnel and aircraft.

---

[1]These definitions are roughly consistent with those tentatively proposed by various AF organizations, although there is not yet a standard terminology. These organizations include AF/ILXX (Lt Col Anthony Dronkers, personal communication, September 1998) and USAFE (U.S. Air Forces Europe).

[2]The LCN expresses the relative effect of an aircraft on pavement; it depends on the aircraft's weight, tire footprint, and tire pressure. See Norman (1996) for a detailed discussion of the LCN, as well as information for all current U.S. aircraft.

[3]This is the current factor in the Air Force's War Resupply Planning Factors database, but it is considered to be quite low.

Category 1

A Category-1 FOL has the attributes of a Category-2 FOL, plus an aircraft arresting system, minimum communications, munitions buildup and storage sites already set up, and three days' worth of prepositioned munitions.

A given airfield may not fit cleanly into one of these categories, and further analysis may show that a more cost-effective arrangement may require that some resources be positioned differently (e.g., re-serving expensive munitions such as missiles to be transported to any FOL). For the purposes of analysis, however, these categories let us consider options of prepositioning or deploying specific sets of re-sources.

SUPPLYING THE DIFFERENCE: FSLs AND CONUS

For each category of FOL, the resources not prepositioned must be supplied during execution to ensure that the supported force meets sortie-generation requirements. The options we consider are an FSL, a logistics operation near the theater of operations, or supply from the CONUS. Our description of the FSL is intentionally vague be-cause there are numerous options for such facilities, ranging from active U.S. bases with airlift support, storage facilities, and repair ca-pability, to simple cold-storage warehouses with minimal mainte-nance. For some of these activities, an FSL could be on a ship that can move to a theater where a crisis is brewing. Much of our ongoing and future work is aimed at informing decisions determining the scope and positioning of FSLs. In this report, we will assume that they are storage facilities from which equipment can be pulled and transported to an FOL.

Chapter Four

EXPEDITIONARY DEPLOYMENT PERFORMANCE: PROTOTYPE ANALYSIS

With the definition of FOLs and the functions of FSLs, we can begin to address how to supply the requirements for each category of FOL, how long each option takes, and roughly estimate the costs for different options. Options are analyzed by the third model component ("TradeMaster"), which uses the outputs from the employment-driven models to compute the values of some of our metrics.[1] In this chapter, we present a prototype analysis that gives the flavor of the analysis that is needed to make strategic infrastructure decisions. We first discuss briefly the scenario we use for this analysis and the metrics for measuring the deployment performance.

EAF SCENARIOS

As described in Chapter Two, our analytic method uses employment scenarios to derive logistics requirements. In this analysis, we give primary attention to a scenario that places heavy demands on those commodities (munitions, POL support, unit maintenance equipment, vehicles, and shelters)[2] that account for most of the support footprint. Although we treat only this one scenario in detail (together

---

[1]There is a TradeMaster model for each of the commodity models. Although each shares a common structure, this structure is modified for each commodity.

[2]In this report, we have focused on munitions and POL support, with some attention to vehicles and shelter. We have also estimated the footprint of maintenance equipment from the maintenance models and for other resources such as communications equipment and medical facilities from the deployment list (TPFDD) of a recent AEF deployment to SWA.

with a small excursion), our models can easily compute requirements and extend this analysis for other scenarios.

The scenario elements that determine the requirements for the major commodities are the number of aircraft and their types (MDS), their sortie rates, their missions (which determine the munitions they carry), and their munitions expenditure rates. The requirement models' key outputs are the people, equipment, and consumables needed. In this analysis, we do not use the personnel requirements, and the equipment and consumables requirements are aggregated for most purposes into costs and gross weight (which can be converted into airlift requirements by using accepted planning factors).

The example scenario illustrated here is heavily influenced by operations in SWA, primarily because attention has been focused on quickly mounting combat operations to that region from CONUS to reduce burdensome repetitive temporary deployments.[3] Therefore, the aircraft, missions, and sortie rates are taken from CENTAF experience.

The basic ASETF in the analyses below consists of

- 12 F-15Cs for air superiority

- 12 F-15Es for ground attack with GBU-10s (2000-lb laser-guided bombs)

- 12 F-16CJs for SEAD (Suppression of Enemy Air Defense) missions.

In our baseline scenarios, these aircraft execute 80 sorties per day (utilization rates of 2.3, 2.3, and 2.0, respectively, assuming all aircraft are always fully mission capable).[4] We consider only materiel required to carry out the first seven days of operations.[5]

---

[3]An example is deployments such as Phoenix Scorpion in 1997, which was a response to Iraqi troop maneuvers near the Kuwaiti border, not to the rotations for enforcement of the various no-fly zones.

[4]This is a demanding scenario, and some have questioned whether such a small force could sustain this optempo for even seven days.

[5]Seven days has emerged as a canonical planning parameter for the initial operation. Clearly, if combat operations are initiated and extended beyond seven days, daily resupply will be a necessity.

PERFORMANCE METRICS

In comparing the performance of infrastructure components both individually and in different configurations, five metrics are of primary interest: timeline, deployment footprint (equipment and people), cost, flexibility, and risk. Our analytic method provides quantitative treatment of the first three, which will be described in more detail below.

Unfortunately, risk and flexibility are more difficult to quantify. One aspect of risk is the probability of having access to an FOL or FSL when a crisis arises. Access can be denied either politically or by military action (a concern especially on the Korean peninsula for FOLs near Seoul). Flexibility is also important: centrally held material can be "swung" more easily to various conflicts than material prepositioned at an FOL. The latter option essentially bets that the conflict will occur within effective range of the FOL.

Risk and flexibility depend heavily on aspects of the global security environment such as where vital U.S. interests are, how greatly they are threatened, and the friendliness of foreign nations that could provide FOLs or space for FSLs. RAND is examining those issues,[6] and in the ongoing work on the integrating model mentioned in Chapter Two we are considering possibilities for quantifying risk and flexibility, but for now decisionmakers must judge the quantitative tradeoffs provided by the logistics modeling with the subjective factors of risk and flexibility.

SCENARIO DEPLOYMENT PERFORMANCE

Figure 4.1 displays the estimates made with the employment-driven and TradeMaster models for six configurations of FOLs and FSLs or CONUS support (each of three categories of FOL in combination with the two options for supplying the remainder). The metrics are displayed in a single figure so that comparison is easier with other configurations and policy or technology options, one of which will be treated later in this report for contrast. We now describe each part of Figure 4.1 for our chosen scenario.

---

[6]James Wendt (1998 unpublished research).

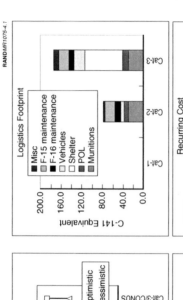

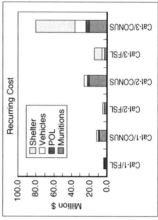

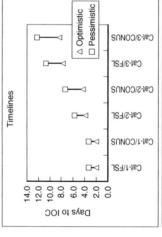

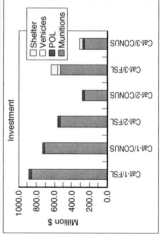

Figure 4.1—Metrics for GBU-10 Case

Timelines to Deploy to Categories of FOL

Each commodity's TradeMaster begins with a list of activities that must be done to set up the process that provides the commodity— deploying personnel from CONUS, moving equipment to the FOL if it is not prepositioned, setting up the equipment, and providing enough support to begin operations. In the TradeMaster, each task requires a deterministic time to complete:[7] some of these times are computed by the requirements models (e.g., the time to set up POL storage facilities or to build the first load of munitions), whereas others are derived from expert judgment or as an assumption (e.g., 22 hours to deploy personnel from CONUS to the FOL). TradeMaster attempts to give an estimate of variability: to a selected set of tasks it adds a (subjective) increment to get a "pessimistic" timeline in addition to the more "optimistic" one.

In some cases, activities can be done in parallel; for example, we assume that, if equipment needs to be moved to the FOL, movement and unloading can be done before the unit personnel arrive. This assumes that the advance teams and particularly the Tanker Airlift Control Element (TALCE) will unload any airlifters in an organized fashion so that incoming personnel can quickly find and move their own equipment. This critical assumption allows us to make a first approximation in integrating the output of the various commodity models: we add the times required to unload the airlift (subject to the MOG constraint) and then take the maximum of that time and all of the other times to set up the various commodity processes and produce the first sortie. This is a rough estimate of the optimistic Initial Operating Capability (IOC). For the pessimistic IOC, we use a similar method on the individual pessimistic IOCs for each commodity and its unloading.

The timelines will also be affected by whether the resources that are not prepositioned are brought from an FSL or from CONUS, and whether in the former case they are moved by airlift or by truck. For

---

[7]As noted above, these models are deterministic because they are intended to be used for strategic planning, not for analyzing specific execution situations.

our purposes, we assume that airlift is always used and that the FSL is four hours by air from the FOL.[8]

The results of the timeline analysis for the three categories of FOL are shown in the upper left-hand panel of Figure 4.1.

The optimistic time to set up a Category-1 FOL is just under two days, even though most equipment is prepositioned. It is primarily driven by the time to deploy the people from CONUS and the time required to set up munitions and fuel-storage facilities. (We have assumed that U.S. forces must set up temporary fuel storage on a prepared site so that fuel for U.S. aircraft can have additives added independently of an ally's fuel.)

For the other options, the times are primarily driven by the MOG and by the assumption of C-141s as the transport aircraft.[9] The difference in timeline between CONUS and an FSL is minimal because the bottleneck is in unloading.[10] For Category-3 FOLs, the primary time driver is unloading the bulky Harvest Falcon package (and setting it up requires 4.6 days with a dedicated 150-person crew in a temperate climate).

In summary, meeting the 48-hour timeline will be virtually impossible with current processes and equipment unless most equipment is prepositioned, and even then the timeline is extremely tight.

Deployment Footprint

We define the deployment footprint as the amount of materiel that must be moved to the FOL for operations to commence. It is derived from the model outputs: the model computes the amounts of equipment and vehicles needed for each commodity and then con-

---

[8]Analysis of an actual situation would require use of real flight times. For example, planners for the Pacific theater would need to use substantially longer times because of the distances between bases.

[9]We use C-141s because these would have been the airlifters used to deploy ASETFs over the first years in which the EAF concept was implemented. They will be out of the force by 2006, when the C-17 will take on this role (but note that the C-17 fleet is much smaller, although its cargo capacity is greater).

[10]This does not take into account the much more demanding air bridge (tankers, etc.) that must be in place to use airlift from CONUS.

verts them to airlift requirements using standard planning factors for each selected aircraft (raw short tons could be used as well).[11] The upper right-hand panel of Figure 4.1 shows the initial airlift requirements for the three categories of FOL (i.e., the amount of airlift required to get the FOL operating).[12]

Peacetime Cost Estimates

Although transportation and materiel costs may properly be ignored when a crisis looms, current fiscal concerns require that part of the evaluation of any set of options include the peacetime costs of setting up a given configuration ("investment") and the costs of operating the system ("recurring"). For example, a Category-1 FOL will require prepositioning of three days' worth of munitions, munitions assembly equipment, and POL storage and distribution equipment. Then the equipment must be maintained for use and be activated for ASETF exercises. If the munitions are to be stored at an FSL for transport to a Category-2 FOL, the FSL must contain enough sets of equipment to cover several ASETF operations in its area.

There are two major omissions from the investment cost:

- The facilities cost for building FSLs or constructing new FOLs. Such costs could be considerable, but FOLs in a theater of interest may be provided by the FOLs of a host country's air force (e.g., Prince Sultan Air Base in Saudi Arabia) at no cost. If an FSL is placed in Europe, it may adapt existing facilities such as Ramstein or Spangdahlem in Germany, the Sanem WRM storage facility in Luxembourg, or Lakenheath in the UK. However, construction in less-developed areas such as at bases on Diego Garcia or in Alaska may be quite expensive. Because these costs

---

[11]The actual computations are a hybrid. For most equipment, we compute the weight in short tons and divide by the capacity of the aircraft that is used for airlift planning purposes. For some bulky equipment, we also use the area taken up to correct the computation or, in some cases, the pallet positions required. The different measures are usually quite close.

[12]The airlift for shelter assumes that this is the housekeeping set (basic billeting, kitchen, power, water, sanitary facilities) in the Harvest Falcon bare-base billeting package.

are so dependent on the actual configuration under considera-
tion, we defer them for this prototype analysis.

- Some of the equipment and consumable costs could be sunk.
For example, the Air Force probably has enough GBU-10s to
preposition them in almost every conceivable configuration of
FOLs and FSLs without buying more. Similarly, there is considerable excess of many kinds of equipment available for storing at
FSLs or FOLs as a result of the recent downsizing.[13] However,
new munitions such as the new standoff munitions coming into
the inventory will need to be bought for any prepositioned storage. For this analysis, we present the total purchase price without considering the sunk costs.

The lower left-hand panel in Figure 4.1 compares investment costs
for our canonical scenario: a 36-ship ASETF carrying out ground
attack with GBU-10s, with moderate rates for missile expenditures.
The configurations are two regions, five FOLs per region (any one of
which might have to support the 36-ship ASETF), and two simultane-
ous ASETF operations (i.e., each central stock location, if any, must
be prepared to support two ASETFs).[14]

As expected, providing for five Category-1 FOLs per region is expensive, and munitions are by far the greatest cost (although recall that
only three days' worth of munitions are prepositioned at each FOL).
Drawing materiel back from the FOLs decreases the cost, increases
flexibility, and (may) decrease risk because each FSL requires only
two sets of equipment. However, airlift requirements increase.

As noted, recurring costs have two components. First is the transportation cost for exercising ASETF deployments. Without a periodic
schedule of exercises, the threat from an ASETF will not be credible
to adversaries. (Of course, the schedule and scope of exercises is a
decision variable.) The airlift cases we describe here use the
Transportation Working Capital Fund (TWCF) price for airlift time.[15]

---

[13]However, aggressive actions to shed excess may have depleted this source.

[14]Each FSL has two sets of equipment. If the materiel is supplied from CONUS,
CONUS needs only two sets total.

[15]This is the rate that the Air Force pays to the U.S. TRANSCOM for peacetime ship-
ments (formerly Defense Business Operating Fund-Transportation or DBOF-T).

The second recurring cost is for storage operations—the cost for having basic security and active maintenance of stored materiel.[16] In the case of vehicles and shelters, we include a 10 percent charge to refurbish the materiel after each exercise, and for options where vehicles are rented we estimate a rental charge of 10 percent of the total cost.

The lower right-hand panel of Figure 4.1 shows our estimates of the recurring costs for these four commodities for the FOL configurations we examine. These recurring costs show a different pattern. The Category-3 FOLs supported from CONUS are very expensive to operate, primarily because of the large costs of transporting munitions and the Harvest Falcon sets twice a year for exercises.[17]

Looking at Figure 4.1 as a whole, we can see that Category-1 FOLs give the fastest response but at high investment costs; Category-2 FOLs have longer response time but at lower investment costs; and FSLs have higher investment costs than stockpiling in CONUS but have lower recurring costs because of the shorter flying time to FOLs.[18] This is an example of the kinds of tradeoffs that need to be considered in designing a strategic support infrastructure: fast response can be purchased by investing in Category-1 FOLs, but this

---

[16]This cost is currently the softest of all of our inputs. It is based on a number of people for a given volume of materiel. However, because the successful deployment of ASETFs will depend on the immediate usability of equipment at FSLs and FOLs, the storage and maintenance policies for prepositioned equipment need to be carefully formulated and rigorously enforced, especially if contractors do some or all of the work. Expeditionary forces will fail if they fall-in on equipment that needs extensive maintenance work before it is serviceable: there simply is not enough time to make major repairs on support equipment and still maintain the credible threat that may be required. In our interviews with Air Force personnel who have taken part in expeditionary operations, this issue has been of serious concern. See General Accounting Office (1998) for more details.

[17]Many of these costs, however, may not have to be borne by the Air Force alone. In some cases, host-nation support may provide some of the services needed. There is also the possibility of sharing facilities with other services: transportation hubs would be useful to the Navy and Army as well. And some USAF and/or Army "FSLs" could even be ship-based, protected by a carrier battle group and supplied by the battle group's resupply pipeline.

[18]Whereas the number of initial airlift is equal for FSL and CONUS options, the airlift for the latter option is strategic (intertheater) airlift—a global asset, as opposed to a theater asset.

may be wasteful if longer timelines can be tolerated because of lower threat or less-critical U.S. interests.

EFFECTS OF DIFFERENT TECHNOLOGIES ON DEPLOYMENT PERFORMANCE

We can use our modeling approach to compare technologies and policies for carrying out the required mission or providing the needed support. In this section, we examine the results of replacing the GBU-10 with the Small Bomb System (SBS), a 250-lb bomb that is designated to be used against approximately 70 percent of the targets that are appropriate for the GBU-10. The small bomb is much lighter than the GBU-10, and each F-15E can carry six small bombs versus two GBU-10s. Thus, the same number of bombs can be dropped in 14 sorties as in the 28 sorties using GBU-10s. This also reduces POL requirements and, with the right scheduling of sorties, refueler requirements. (We have assumed that the same number of air-escort and SEAD missions are flown.) The SBS is considerably more expensive than GBU-10s, however, and the effect on the costs of prepositioning options is uncertain because the bombs cost more but the reduced sorties mean lower missile expenditures.[19] Figure 4.2 compares the alternative support options using the three metrics if the SBS substitutes for the GBU-10.

The general pattern of each metric seems similar in this case, but closer comparison shows significant differences between the two cases.

---

[19]The SBS is only under test and has not been procured. The costs shown here are therefore dollars that must be programmed and expended, unlike the costs for the GBU-10, which are largely sunk. A more detailed analysis would need to compare the two sets of costs with the GBU-10 costs subtracted.

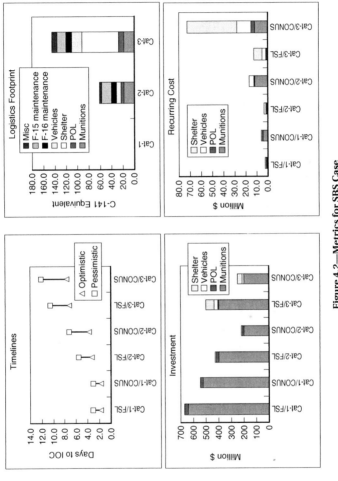

Figure 4.2—Metrics for SBS Case

The SBS option as presented here does degrade the startup performance slightly, because the increased bomb load per sortie requires more bomb buildup work per flight (and hence more for the first flight). The SBS is supposed to be able to be shipped in a full-up configuration, however, so it may be feasible to prebuild the rounds on strategic warning at a storage site and reduce the time to IOC. As expected, the initial airlift required is somewhat smaller, although the weight of munitions-handling equipment is still significant. Finally, the investment and recurring costs are lower for the SBS option. The investment decreases because of fewer expenditures of air-to-air missiles—the mission can be carried out with fewer sorties. Recurring costs are reduced because airlift needed to transport SBSs for exercises is less. Note that this discussion has implicitly assumed that rapid transportation is available for movement of munitions to an FOL when the munitions are stored in an FSL or in CONUS.

One reviewer asked whether a deploying force might not elect to bring both SBS and older rounds, thereby not saving any deployment footprint. Our analyses have been predicated on specifying which munitions will be used and evaluating the deployment performance. Greater flexibility, in terms of a mix of munitions, must be paid for with more transportation or more prepositioning.

CONCLUSIONS

After looking at the current force structure and its support processes, our analysis leads to several conclusions concerning FOLs and their support.

- To get close to the 48-hour deadline from execution order to placing the first bombs on target, ASETFs must deploy to Category-1 FOLs. Further, given that a flight halfway around the world takes approximately 20 hours, pushing the timeline below 48 hours will require having people deployed, or materiel at an advanced state of preparation at the FOL, or both.

- Equipping several FOLs from scratch would be expensive. Although much of the cost for current processes might well be sunk, maintenance and storage costs will still have to be paid. Anecdotal accounts of current (non-urgent) deployments to SWA indicate that maintenance arrangements do not keep equipment ready for immediate use, implying that these costs might be larger than are paid now. Further, future munitions and improved support equipment would have to be bought for the FOLs.

- FSLs provide a compromise in cost between prepositioning at FOLs and deploying everything from CONUS.[20] FSLs have little effect on the timeline for initial capability, but they do avoid the necessity of a tanker air bridge for the extra strategic lift. Further, this strategic lift then becomes available for further deployments, which may be needed if the crisis is not resolved.

- Category-2 FOLs represent another compromise in cost and timeline. However, to deploy to a Category-2 FOL would take between two and three days to unload the airlift and about the same amount of time to set up munitions and POL storage, so increased ramp space would not significantly speed up the deployment process because operations could not commence until the setup was completed.[21] Plus, arrangements for rental vehicles, medical facilities, and the like would probably require some time to finalize unless complete preparations had been made in advance.

- Category-3 FOLs are not useful as FOLs for very quick crisis response, given the time required to unload airlift and set up the processes. However, this is a function of the current processes, and the timeline estimated here is for a stressing scenario. A less-stressing scenario or a humanitarian operation might well be feasible from such a Category-3 FOL within the 48-hour timeline.

---

[20]However, much of the difference in recurring costs arises because of the expense of running exercises from CONUS and the form of the exercises.

[21]This assumes that POL and munitions troops and equipment arrive early in the deployment sequence.

Chapter Five

ANALYZING OPTIONS FOR EXPEDITIONARY ACS

The current security environment, with its requirement for many small operations and the pressures to reduce large overseas presence, has given rise to the concepts of the Expeditionary Aerospace Force and halt-phase operations in MTWs. As originally envisioned, the EAF would consist of air and space forces that could be tailored to emerging crises, deployed rapidly to the required location, and be prepared to execute operations immediately (within 48 hours of departure from home), even from an FOL with virtually no infrastructure beyond a usable runway. The prototype analyses in Chapter Four indicate that this is not possible:

> With today's support processes, policies, and technologies, deploying even a modest fighter-based combat force to a bare base will require several days of development before the FOL can sustain a high-flying tempo.

This finding does not mean that achieving the 48-hour operational goal is impossible. The analyses here have indicated the goal can be met by developing a strategic theater infrastructure with the judicious prepositioning of equipment, materiel, and facilities. Although such prepositioning would require a substantial investment and the assumption of increased political and military risk, it does not require the development of new technologies or support processes. But it does require hard thinking about the nature of the threat and the level of U.S. interests involved to ensure that such investment is worth the cost.[1]

There are, however, other options. One is to decide that the threat or the criticality of U.S. interests does not require 48-hour response; in this case, more austere FOLs may be suitable for the longer timeline. There are also operational options that might have more palatable support requirements. For some threats, bomber operations may be as effective as using fighters in air-to-ground attacks. Although a first strike can be carried out when launching from CONUS, repeated strikes would probably have to take place from a closer FOL. However, the greater range of bombers means fewer FOLs would be needed for bombers than for fighters. Such FOLs could be carefully positioned where governments were reliably friendly to the United States (e.g., Diego Garcia Island, which is controlled by the United Kingdom).[2]

Another alternative is to change the current processes. As also illustrated in Chapter Four, technology improvements in a key area can shift some of these options. For example, lighter munitions (assuming that they are equally effective) do not require prepositioning because they can be transported more easily than current munitions. Similarly, other analyses have indicated that aerospace ground equipment (AGE) is the largest component of unit maintenance equipment.[3] Finally, the great weight of the current bare-base

---

[1]For example, in designing a system of FOLs and FSLs, we must also consider how operations other than AEFs will be supported. Force requirements for the current set of MTWs that are used for U.S. conflict planning are set out in the *Defense Planning Guidance* (DPG) (Department of Defense, 1998). Considering both the in-place forces and the first week of fighting, a rough estimate of the basing infrastructure required is two or three Category-1 bases plus one or two Category-2 bases (assuming that such bases can be prepared in about five days) in each theater to meet the timeline and bed down the incoming force. Recent work cited in Chapter One on effect-based operations (or halt-phase operations) shows that such operations require deploying a substantially larger force within a few days rather than the several weeks envisioned in the DPG. Given the results in Chapter Three, this would require four to five Category-1 bases in the theater, with another one or two Category-2 bases available at the end of the first week.

[2]Work along these lines has been done notably by AF/ILXB (Maj Barr, AF/ILXB, personal communication, 1998).

[3]The Air Force has several efforts that address this issue, but coordination to date seems to have been relatively informal. These include research overseen by the AEF Battle Lab to develop a combination generator/air conditioner unit (briefing by LTC

package is one reason why deployments to Category-3 FOLs take so long. There is considerable potential here for commercial alternatives both in shelter and perhaps security technology.[4] However, any of these options requires a delay for research and procurement.

We assert that the long-term support issues raised in this report about FOLs, FSLs, and their locations and equipage require detailed analyses that allow comparison over a wide range of scenarios. Such analyses can be made along the lines of the prototype analyses here. Further, such analyses need to be carried out with a *strategic* perspective, one that views the entire support structure, both inside and outside CONUS, as a *system* of global support. As described more fully in a companion report on planning for the EAF,[5] many such global decisions need to be made centrally in order to make consistent use of scarce resources. Resulting decisions need to be revisited on a regular basis as the global political situation changes and as technology changes the capabilities of the Air Force.

Jeff Neuber on AEF Battle Lab initiatives, December 2 1998), the Aerospace Ground Support Equipment Working Group in AF/ILMM, and a comprehensive program at the Air Force Research Laboratory that is developing modular ground equipment that can be reconfigured for different aircraft types.

[4]The Airbase Systems Command at Eglin Air Force Base is overseeing research on new shelter technology and other aspects of bare-base infrastructure. CENTAF has investigated alternative security barrier technologies.

[5]See Tripp et al., 1999.

MODEL OUTPUT FOR PROTOTYPE CASES

These tables tabulate the numerical values that underlie Figures 4.1 and 4.2.

Table A.1

Spinup Time for GBU-10 Case
(days)

	Cat-1 FSL	Cat-1 CONUS	Cat-2 FSL	Cat-2 CONUS	Cat-3 FSL	Cat-3 CONUS
Optimistic	1.9	1.9	3.8	4.4	7.7	8.3
Pessimistic	3.3	3.3	5.8	7.4	10.7	12.3

Table A.2

Investment Cost for GBU-10 Case
(million $)

	Cat-1 FSL	Cat-1 CONUS	Cat-2 FSL	Cat-2 CONUS	Cat-3 FSL	Cat-3 CONUS
Munitions	854.5	702.4	524.3	262.2	524.3	262.2
POL	28.9	28.9	24.4	23.0	11.5	5.8
Vehicles	0.0	0.0	8.0	4.0	29.4	14.7
Shelter	0.0	0.0	0.0	0.0	68.1	34.0

Table A.3

Recurring Cost for GBU-10 Case
(million $)

	Cat-1 FSL	Cat-1 CONUS	Cat-2 FSL	Cat-2 CONUS	Cat-3 FSL	Cat-3 CONUS
Munitions	2.5	9.4	2.5	19.8	2.5	19.8
POL	0.4	0.4	0.5	1.8	0.5	4.2
Vehicles	1.5	1.5	1.8	4.3	2.8	12.6
Shelter	0.0	0.0	0.0	0.0	8.5	43.8

Table A.4

Footprint for GBU-10 Case
(C-141 equivalents)

	Cat-1	Cat-2	Cat-3
Munitions	0.0	30.8	30.8
POL	0.0	6.8	10.7
Shelter	0.0	0.0	72.0
Vehicles	0.0	7.3	20.6
F-16 maintenance	0.0	11.2	11.2
F-15 maintenance	0.0	18.6	18.6
Miscellaneous	0.0	4.6	9.5

Table A.5

Spinup Time for SBS Case
(days)

	Cat-1 FSL	Cat-1 CONUS	Cat-2 FSL	Cat-2 CONUS	Cat-3 FSL	Cat-3 CONUS
Optimistic	1.8	1.8	3.4	4.0	7.3	7.9
Pessimistic	3.2	3.2	5.7	7.3	10.7	12.3

Table A.6

Investment Cost for SBS Case
(million $)

	Cat-1 FSL	Cat-1 CONUS	Cat-2 FSL	Cat-2 CONUS	Cat-3 FSL	Cat-3 CONUS
Munitions	640.0	522.9	396.5	198.3	396.5	198.3
POL	26.1	26.1	22.4	21.2	10.4	5.2
Vehicles	0.0	0.0	8.0	4.0	29.4	14.7
Shelter	0.0	0.0	0.0	0.0	68.1	34.0

Table A.7

Recurring Cost for SBS Case
(million $)

	Cat-1 FSL	Cat-1 CONUS	Cat-2 FSL	Cat-2 CONUS	Cat-3 FSL	Cat-3 CONUS
Munitions	1.4	4.4	1.5	11.6	1.5	11.6
POL	0.4	0.4	0.4	1.5	0.5	3.6
Vehicles	1.5	1.5	1.8	4.3	2.8	12.6
Shelter	0.0	0.0	0.0	0.0	8.5	43.8

Table A.8

Footprint for SBS Case
(C-141 equivalents)

	Cat-1	Cat-2	Cat-3
Munitions	0.0	21.1	21.1
POL	0.0	6.3	9.4
Shelter	0.0	0.0	72.0
Vehicles	0.0	7.3	20.6
F-16 maintenance	0.0	11.2	11.2
F-15 maintenance	0.0	18.6	18.6
Miscellaneous	0.0	4.6	9.5

BIBLIOGRAPHY

Air Force Instruction 10-400, *Aerospace Expeditionary Force Planning*, draft, June 1999.

Davis, Richard G., *Immediate Reach, Immediate Power: The Air Expeditionary Force and American Power Projection in the Post Cold War Era*, Air Force History and Museums Program, Washington, D.C., 1998.

Department of Defense, *FY 1998–2003 Defense Planning Guidance*, Washington, D.C., 1998 (classified report).

Fuchs, Ronald P., *Report of the United States Air Force Scientific Advisory Board Committee on United States Air Force Expeditionary Forces*, Department of the Air Force, Washington, D.C., 1997.

General Accounting Office, *Military Prepositioning: Army and Air Force Programs Need to Be Reassessed*, GAO/NSIAD-99-6, Washington, D.C., November 1998.

Hosek, James R., and Mark Totten, *Does Perstempo Hurt Reenlistment? The Effect of Long or Hostile Perstempo on Reenlistment*, RAND, MR-990-OSD, 1998.

Norman, Bill, *Weight Bearing Capacity Evaluations*, AFIF Data Dictionary, Amendment 2, National Imagery and Mapping Agency, 1996.

Richter, Paul, "The Tough Job of Keeping Soldiers Ready for War," *Los Angeles Times*, November 22, 1998.

Richter, Paul, "Buildup in Gulf Costly: Expenses, Stress Surge for Military," *Los Angeles Times*, November 17, 1998.

Ryan, General Michael E., *Evolving to an Expeditionary Aerospace Force*, Commander's NOTAM 98-4, Washington, D.C., July 28, 1998.

Tripp, Robert S., Lionel A. Galway, Paul S. Killingsworth, Eric Peltz, Timothy L. Ramey, and John G. Drew, *Supporting Expeditionary Aerospace Forces: An Integrated Strategic Agile Combat Support Planning Framework*, RAND, MR-1056-AF, 1999.

Washington Post, "Saudis slow to back US in Iraqi crisis," February 3, 1998, p. A01.

Williams, Matthew, "Plea for help (from the Air Force secretary and the Chief of Staff): Better pay, bigger budgets called key to fixing readiness woes," *Air Force Times*, September 28, 1998.